CAPTURING SCENTS

A Concise Guide to Essential Oil and Aromatics Extraction

Dr. Scott A. Johnson

Capturing scents: A concise guide to essential oil and aromatics extraction / Scott A. Johnson

Cover design: Scott A. Johnson
Cover Copyright: © Scott A. Johnson 2025

ISBN-13: 979-8988720645
Published by Scott A. Johnson Professional Writing Services, LLC: Orem, UT

Discover more books by Scott A. Johnson at authorscott.com/shop/

Table of Contents

CHAPTER ONE: The Evolution of Extraction—From Ancient Techniques to Modern Innovations

The civilizations of old learned to tap into the rich, aromatic bounty of nature by developing crude methods of extraction to harness volatile organic compounds for a wide range of uses, from medicine and cosmetics to perfumery and embalming. Our ancestors, deeply connected to the plant world, discovered that some plants held unique properties that benefited health, beauty, and cultural life.

The earliest civilizations, devoid of the complicated equipment at their disposal today, relied on ingenuity and observation to devise means of capturing the essence of plants. Traditional methods most commonly used included infusions, maceration, expression, enfleurage, and crude forms of distillation-all forerunners of the modern techniques of extraction at our disposal today.

Infusions and decoctions involved soaking plant materials in liquid, usually water or oil, to extract the active principles from them. The liquid used in most cases was related to the purpose for which the material was being prepared; for example, medicinal teas usually used water, while cosmetic products used oils. These procedures, though simple, showed an understanding of solubility and thus, in their own way, the ancients were novice chemists.

Another common method was maceration—a method involving the crushing or grinding of plant materials before steeping them. This process of extraction was effective because it increased the surface area of the plant material exposed to the solvent. Evidence of this practice can be found in ancient Egyptian texts, where oils infused with the essences of various plants were used not only in religious rituals but also in daily life.

Another widely used technique was expression, or the mechanical expression of botanicals, primarily peels of citrus fruits. This process extracted fragrant essential oils without the application of heat, thus preserving the sensitive makeup of the volatiles. History relates that this process was utilized to produce exquisite perfumes and flavorings, demonstrating the linkage of nature to the new field of aromatics.

One of the most interesting extraction techniques was enfleurage: placing fragrant flowers in odorless fats so that, over a period of time, these absorbed the flowers' essential oils. This was an extremely laborious process and was used mostly for delicate flowers that could not be exposed to heat without destruction, such as jasmine and tuberose. The fragrant fats were then treated to eliminate the aromatic constituents, an early example of an application of chemistry and sensitivity toward the requirement of preserving the integrity of plant material.

The development of various, rudimentary forms of distillation allowed ancient cultures the ability to heat plant materials and capture condensed vapors. These early techniques were not as sophisticated as today's methods of distillation but did begin to uncover the complexity of natural fragrances and aromas and thus laid the foundation for the modern essential oil industry.

It is both a humbling and inspiring exercise to think through how these historical practices have evolved to give way to sophisticated methods that help in better extraction of essential oils and aromatic extracts. The tapestry of human history is so intently interwoven with the use of plants that our journey to understand and optimize the methods of extraction has been propelled by the desire to maximize the beneficial properties of these natural compounds.

Today, scientific research and technological development remain the backbone in the improvement of methods of extraction. In fact, with more research and practitioners looking to achieve better yields, aroma profiles, and selective extraction of these volatile aromatic compounds, a renaissance period for aromatic extraction is currently transpiring. Since each of these techniques presents its own set of advantages and challenges, while some result in the production of pure essential oils, others yield aromatic extracts capable of taking along a broader spectrum of secondary metabolites.

Each method of extraction has its various strengths and weaknesses, which accordingly affect the ultimate quality of either the essential oil or the aromatic extract. Whereas steam distillation is widely adopted industrially owing to its efficacy and historical use, for some floral scents that are extremely delicate, methods such as solvent extraction or even enfleurage would better maintain the complexity in their aroma profile.

One of the major challenges in extraction is finding the right balance among efficiency, sustainability, and product quality. As awareness related to the environment increases among consumers, a growing interest is evidenced in greener extraction methods (e.g., CO_2 supercritical extraction), which minimize waste and energy consumption while maximizing yield and quality. This demand is driving research and innovation, and new approaches are surfacing that favor sustainability without compromising quality.

Besides, continuous development within analytical chemistry enabled the improvement of extraction processes and deepened knowledge about the chemical composition of essential oils and extracts. Modern techniques, such as gas chromatography-mass spectrometry, provide detailed profiles of the various compounds present, enabling producers to tailor their extraction methods further in order to capture specific aromatic compounds for particular applications.

Over recent years, researchers have considered several variables to make extraction efficient. Temperature, pressure, and the time of extraction have become crucial variables in studying the efficiency of extraction. From these variables, a balance exists where yield and quality can both be optimized with the highest possible standards that customers would like to see.

As we begin to explore the world of essential oils and aromatic extracts in more depth, so too must we consider-aside from the technical aspects of extraction-the cultural significance and historical context in which these practices are set. Indeed, ancestral techniques continue to echo today and remind us about our ongoing relation with nature and our quest to harness its potential.

In all, the process of essential oil and aromatic extract extraction represents a story with roots tracing as far back as several millennia. From the early days of infusions and enfleurage through to the state-of-the-art techniques we apply today, every phase has played its role of adding knowledge and methods that improved our understanding and use of aroma compounds. By saving the wisdom and art of the ancients and adding to it the genius of modernity, the doors are now opening to a new era in the field of aroma extraction, one that is qualitative, sustainable, and respectful to nature. The sections to follow will delve into each extraction method in detail, along with the chemistry involved, applications of the produced extracts, and recent developments that could change the game. Not only will you have a better understanding of how these aromatic gifts are extracted, but hopefully you'll learn to appreciate the intricate dance among nature, culture, and science that so typifies this exciting field.

CHAPTER TWO: Steam Distillation

Examples of essential oils obtained via this method: lavender, peppermint, tea tree, eucalyptus, oregano, lemongrass, thyme, hinoki, cedarwood, myrtle, copaiba, cardamom, many others.

Steam distillation is one of the most widely used methods for extracting essential oils from plant materials. The process involves the generation of steam, which is passed through the plant material (such as flowers, leaves, or stems) contained in a still. As the steam circulates through the plant, it causes the essential oils within the plant cells to evaporate.

The mixture of steam and vaporized essential oils then travels through a condenser, where it cools and reverts back to liquid form. This liquid is collected in a receiving vessel, consisting of two phases: a layer of essential oil, which floats on top of the water—with the exception of essential oils denser than water, like clove and wintergreen. The essential oil can easily be separated from the water due to the difference in their densities.

Advantages

- Preservation of Volatile Compounds: Steam distillation operates at relatively low temperatures, which helps preserve the delicate volatile compounds in essential oils. This is particularly beneficial for heat-sensitive plant materials.

- High Yield: Steam distillation can produce high yields of essential oils, especially from plant materials that are rich in moisture.

- Efficient Separation: The process allows for the efficient separation of essential oils from water (hydrosol) due to their different densities.
- Suitable for Various Plant Materials: Steam distillation is versatile and can be used to extract essential oils from a wide range of plant materials, including flowers, leaves, and stems.
- Continuous Operation: Steam distillation can be operated continuously, making it suitable for industrial-scale essential oil production.

Disadvantages

- Equipment Requirements: The equipment for steam distillation is more complex and expensive than hydrodistillation, requiring a condenser and a boiler.
- Energy Consumption: The process requires significant amounts of energy to generate steam, which can increase operational costs.
- Water Usage: Steam distillation requires a substantial amount of water, which may be a concern in areas where water is scarce.
- Dependence on Water Quality: The quality of the water used in steam distillation can affect the quality of the essential oil produced.
- Risk of Contamination: If not properly maintained, the equipment can become contaminated, affecting the quality of the essential oils produced.

Overall, steam distillation is a widely used and effective method for extracting essential oils, offering several advantages in terms of preservation of volatile compounds and efficient separation. However, it also presents some challenges related to equipment requirements, energy consumption, and water usage.

CHAPTER THREE: Hydrodistillation

Examples of essential oils obtained via this method: Rose, neroli, chamomile, basil, frankincense, clove, citruses.

Hydrodistillation is a technique used to extract essential oils from plant materials by immersing them in water and then boiling the mixture. In this process, the plant material is placed in a still containing water, which is heated to boiling. The steam generated carries the volatile essential oils from the plant. As the steam rises, it cools and condenses, forming a mixture of water and essential oils that is collected in a receiving vessel.

Advantages

- Gentle Extraction: Hydrodistillation operates at lower temperatures compared to other methods like steam distillation, which helps preserve sensitive compounds in the oils.
- High Yield: This method can often produce higher yields of essential oils, especially from plant materials that are rich in moisture.
- Simple Setup: The equipment for hydrodistillation is relatively straightforward, making it accessible for small-scale producers.

Disadvantages

- Water Solubility: Some compounds may dissolve in water during the process, potentially leading to a loss of valuable aromatic constituents.
- Extended Time: The hydrodistillation process can be time-consuming, requiring longer distillation periods than steam distillation.

- Energy Consumption: Boiling water for extended periods can lead to higher energy costs, particularly in industrial applications.

Overall, hydrodistillation is a reliable method for extracting essential oils, especially when careful monitoring of temperatures and times is practiced, but it may not be optimal for every plant type.

CHAPTER FOUR: Molecular Distillation

Examples of essential oils or aromatic extracts obtained via this method: Sandalwood, citrus oils, hemp, ginger.

Molecular distillation is a specialized separation technique used to extract essential oils and other volatile compounds by utilizing difference in volatility under low pressure. In this method, the plant material is subjected to high temperatures in a vacuum environment, allowing for the separation of essential oils based on their molecular weight and boiling points. Because the process occurs under reduced pressure, temperatures are kept low, minimizing thermal degradation of the sensitive compounds. The vaporized essential oils are then condensed into a liquid, isolating them from any remaining plant material.

Advantages

- Preservation of Quality: The low-temperature operation helps maintain the integrity and quality of heat-sensitive compounds, preserving the aroma and therapeutic properties of the essential oils.
- High Purity: Molecular distillation can yield very pure essential oils with minimal impurities, making it ideal for the production of high-quality extracts.
- Efficient Separation: This method allows for the separation of compounds with very close boiling points, which can be challenging with traditional distillation methods.

Disadvantages

- High Cost: The equipment and operational costs for molecular distillation are significantly higher than conventional distillation methods, limiting its use mostly to large-scale producers.
- Complexity: The process requires skilled operators and sophisticated equipment, which can complicate production.
- Limited Scalability: While it excels in purity, molecular distillation may not be as scalable for large quantities compared to other methods like steam distillation.

In summary, molecular distillation offers advantages in quality and purity for essential oil extraction, but it comes with higher costs and operational complexities.

CHAPTER FIVE: Fractional Distillation

Examples of essential oil fractions obtained via this method: ylang ylang, bergamot, nutmeg, ho wood, furanocoumarin-reduced citruses, frankincense, sandalwood.

Fractional distillation is a technique used to separate essential oil components based on their boiling points. It involves heating the oil in a controlled manner, allowing different fractions to evaporate and condense at specific temperatures. This method is particularly useful for refining essential oils, isolating specific compounds, or removing unwanted constituents.

The process occurs in a fractionating column, where compounds with lower boiling points evaporate first and are collected separately from heavier molecules that condense later. This enables the production of essential oils with enhanced purity, potency, and targeted chemical profiles.

Advantages

- Enhanced Purity: Removes impurities, heavy molecules, and unwanted constituents (e.g., waxes, furanocoumarins).
- Customization: Allows for the isolation of specific components (e.g., high-eugenol clove oil or high-santalol sandalwood oil).
- Reduced Thermal Degradation: Controlled temperature prevents the breakdown of delicate compounds.
- Improved Stability: Eliminates volatile fractions that contribute to oxidation and spoilage.

Disadvantages:

- Complex & Costly: Requires specialized equipment and expertise.
- Potential Loss of Minor Compounds: Some trace elements contributing to the full aroma or therapeutic profile may be lost.
- Not Suitable for All Oils: Some essential oils, like citrus oils, are better extracted via cold pressing or CO_2 extraction.

Fractional distillation is widely used in perfumery, pharmaceuticals, and aromatherapy to refine and enhance essential oil compositions.

CHAPTER SIX: Cold Press Extraction/Expression

Examples of essential oils obtained via this method: orange, lemon, tangerine, grapefruit, lime, bergamot.

Cold press extraction, also known as expression, is a method primarily used for extracting essential oils from citrus fruits like oranges, lemons, and limes. This process involves mechanically pressing the peels of the fruit to release the essential oils contained in the oil sacs. The peels are often subjected to pressure in specialized presses, sometimes with the addition of water to aid in oil release or rasping machines that break the sacs. The resulting mixture of oil, water, and fruit debris is then typically centrifuged or filtered to separate the essential oil from the other components. The technique does not apply heat, which helps to preserve the delicate aromatic compounds and flavors present in the oils. It also means that the resulting essential oil will contain waxes and other nonvolatile compounds.

Advantages

- Retention of Flavor and Aroma: Because there is no heat involved, cold press extraction preserves the natural aroma and flavor of the essential oils, resulting in high-quality, vibrant scents.
- Simplicity: The method is relatively straightforward and does not require complex equipment, making it accessible for smaller producers.
- Environmentally Friendly: Cold pressing uses minimal energy and does not involve solvents or

chemicals, making it an eco-friendly method of extraction.

Disadvantages

- Limited to Certain Fruits: Cold press extraction is mainly suitable for citrus fruits, restricting its application to a narrower range of essential oils.
- Lower Yield: The yield of essential oils from cold pressing can be lower compared to methods like steam distillation.
- Labor Intensive: The manual or semi-automated nature of the process can require more labor, making it less efficient for large-scale production.

In summary, cold press extraction is an effective method for obtaining high-quality essential oils from citrus fruits, with advantages in flavor retention and simplicity but limitations in yield and the range of usable materials.

CHAPTER SEVEN: Microwave-Assisted Extraction

Examples of aromatic extracts obtained via this method: lavender, rosemary, eucalyptus.

Microwave-assisted extraction (MAE) is an innovative technique that uses microwave energy to extract aromatic compounds from plant materials. In this method, microwave radiation heats the solvent and plant cells simultaneously, leading to rapid cell disruption and enhanced diffusion of volatile compounds into the solvent. The microwaves cause the moisture within the plant cells to vaporize, which then carries the essential oils out of the plant material and into the surrounding solvent. This results in a more efficient extraction process compared to traditional methods.

Advantages

- Increased Efficiency: MAE significantly reduces extraction time—often completing the process in minutes—compared to several hours or days for conventional methods.
- Higher Yield: The disruption of plant cell walls allows for a greater release of aromatic compounds, often resulting in higher yields of essential oils or extracts.
- Lower Solvent Usage: MAE requires less solvent, making it a more cost-effective and environmentally friendly option than traditional extraction techniques.
- Enhanced Selectivity: The method allows for better control over extraction conditions, leading

to more selective extraction of specific compounds.

Disadvantages

- Equipment Costs: The initial investment in microwave extraction equipment can be high, which may deter small-scale producers.
- Heat Sensitivity: Although MAE operates under controlled conditions, some heat-sensitive compounds might degrade if not carefully monitored.
- Isomerization: Microwave energy can potentially cause isomerization, which is a change in the spatial arrangement of molecules. This could alter the composition of the essential oil and potentially affect its properties.
- Formation of Artifacts: In some cases, microwave exposure might lead to the formation of chemical artifacts, which are compounds that are not naturally present in the plant but are created during the extraction process. These artifacts could potentially have undesirable effects.
- Limited Applicability: The technique may not be suitable for all plant materials or extraction purposes, requiring additional research for optimization.

In summary, microwave-assisted extraction offers several advantages in efficiency, yield, and solvent usage while presenting challenges related to equipment cost and potential compound degradation or alteration.

CHAPTER EIGHT: Ultrasound-Assisted Extraction

Examples of aromatic extracts obtained via this method: ginger, turmeric.

Ultrasound-assisted extraction (UAE) is a modern extraction technique that utilizes ultrasonic waves to enhance the extraction of aromatic compounds from plant materials. This method involves immersing the plant material in a solvent and applying high-frequency sound waves, which create cavitation bubbles. These bubbles implode, generating shock waves that disrupt cell structures, facilitating the release of essential oils and other volatile compounds into the solvent.

Advantages

- Increased Extraction Efficiency: UAE significantly accelerates the extraction process, often reducing extraction times from hours to minutes, making it time-efficient.
- Higher Yield: The ultrasonic waves promote greater diffusion and penetration of the solvent, often resulting in higher yields of aromatic extracts compared to traditional methods.
- Lower Temperatures: UAE typically operates at lower temperatures, minimizing the risk of heat-induced degradation of sensitive compounds, thereby preserving the quality of the extracts.
- Reduced Solvent Use: This technique can reduce the amount of solvent needed, making it more environmentally friendly and cost-effective.

Disadvantages

- Equipment Costs: The initial investment in ultrasound extraction equipment can be significant, which may limit its use to larger facilities.
- Optimization Needs: UAE parameters (such as frequency, power, and duration) require careful optimization for different materials to achieve the best results.
- Potential Over-Extraction: Incorrect parameters can lead to the extraction of undesirable compounds, such as pigments or waxes, affecting extract purity.

In short, ultrasound-assisted extraction is a powerful method that enhances extraction efficiency and quality, though it requires careful optimization and can involve significant upfront costs.

CHAPTER NINE: Solvent Extraction

Examples of aromatic extracts obtained via this method: jasmine, rose, tuberose, benzoin, osmanthus.

Solvent extraction is a traditional and widely used method for isolating aromatic compounds from plant materials. In this process, the plant material is immersed in a suitable solvent that selectively dissolves the desired volatile compounds. Depending on the solubility and polarity of the target metabolites, various organic solvents can be employed. Common solvents used include ethanol, methanol, acetone, and hexane.

Advantages

- Simplicity and Cost-Effectiveness: Solvent extraction methods are relatively straightforward and economically feasible, making them accessible for both small and large-scale operations.
- Versatility: The technique can be adapted for various aromatic compounds by selecting appropriate solvents, allowing for a broad range of applications.
- Good Yield: It often yields high concentrations of essential oils and other aromatic extracts.

Disadvantages

- Solvent Residue: There is a risk of residual solvents in the final extract, which may require additional purification steps to ensure safety and quality.
- Environmental Concerns: Many organic solvents can be harmful to the environment and human

health, necessitating stringent handling and disposal measures.

- Heat Sensitivity: High temperatures may be involved, which could degrade sensitive compounds.

In summary, while solvent extraction is a versatile and effective method for obtaining aromatic extracts, it carries challenges related to solvent residue and environmental safety.

CHAPTER TEN: Maceration

Examples of aromatic extracts obtained via this method: calendula, jasmine, gardenia, vanilla, frankincense, myrrh.

Maceration is a traditional extraction technique used to obtain aromatic extracts from plant materials. This method involves soaking chopped or crushed plant material in a solvent, usually ethanol, oil, or water, for an extended period. During this time, the solvent penetrates the plant tissues, dissolving the desired aromatic compounds. The mixture is then filtered to separate the solid residue from the liquid extract.

Advantages

- Simplicity: Maceration is a straightforward process that requires minimal equipment and can easily be performed in laboratory or home settings.
- Preservation of Compounds: Since the extractive process occurs at low temperatures, maceration helps preserve sensitive aromatic compounds that might otherwise degrade with higher heat methods.
- Cost-Effective: The method typically requires less initial investment compared to more sophisticated extraction techniques, making it accessible for small-scale producers and hobbyists.
- Quality of Extracts: Macerated extracts are often rich in aroma and flavor, as the gentle process allows for the retention of volatile compounds.

Disadvantages

- Time-Consuming: Maceration can take several hours to days, depending on the desired concentration, making it less efficient than other extraction methods like SFE.

- Low Yield: The yields can be lower compared to more aggressive methods, as not all of the compounds may be fully extracted.

- Solvent Recovery: Recovery of the solvent can be challenging, and residual solvent may remain in the extracts, requiring further purification.

To summarize, maceration is a simple and effective method for extracting aromatic compounds, offering purity and quality, albeit at the cost of time and yield.

CHAPTER ELEVEN: Enfleurage

Examples of aromatic extracts obtained via this method: jasmine, tuberose, rose, gardenia.

Enfleurage is an ancient and labor-intensive method of extracting aromatic compounds from flowers, particularly those with delicate fragrances, such as jasmine and rose. This technique involves placing freshly picked flowers onto a layer of warmed, odorless fat, which absorbs the essential oils over time. The flowers are typically replaced several times to maximize the extraction, after which the fat is treated with alcohol to separate the aromatic compounds from the carrier fat.

Advantages

- Low Temperature: Enfleurage operates at ambient conditions, preventing the degradation of volatile aromatic compounds that are sensitive to heat.
- Gentle Extraction: This method is suitable for delicate flowers whose fragrances could be lost or altered through steam distillation or solvent extraction.
- High-Quality Extracts: The process yields a high-quality absolute with rich and complex aromas, making it highly valued in perfumery.

Disadvantages

- Cost and Labor-Intensive: Enfleurage is time-consuming and requires substantial manual labor, making it economically unviable for large-scale commercial applications.

- Limited to Certain Flowers: It is primarily effective for specific flowers with high oil content, thus not applicable for all aromatic plants.
- Fat Recovery: The extraction process involves using fats, which need to be treated with alcohol to recover the aromatic material, adding another step to the procedure.

To recap, while enfleurage is a valuable method for capturing the essence of delicate florals, its labor intensity and specific applicability present significant limitations compared to more modern extraction techniques.

CHAPTER TWELVE: Subcritical Fluid Extraction

Examples of aromatic extracts obtained via this method: cinnamon, clove, other spices.

Subcritical fluid extraction is an innovative extraction method that utilizes subcritical fluids—liquids held at temperatures and pressures below their critical point—to selectively extract aromatic compounds from plant materials. Commonly, water and carbon dioxide are used as the subcritical fluids. In this process, the fluid penetrates the plant matrix, dissolving and carrying away the desired compounds without the high temperatures associated with traditional extraction methods.

Advantages

- Efficiency: Subcritical fluid extraction typically yields higher extraction efficiency and speed compared to conventional methods, making it suitable for large-scale production.

- Selectivity: The method can be fine-tuned to target specific compounds, allowing for the extraction of desired aromatic profiles while leaving behind undesired constituents.

- Environmentally Friendly: Using carbon dioxide, a non-toxic and recyclable solvent, subcritical fluid extraction presents a greener alternative to chemical solvents, reducing environmental impact.

- Preservation of Quality: The low-temperature operation minimizes thermal degradation, preserving the integrity of sensitive aromatic compounds.

Disadvantages

- High Initial Costs: The equipment required for subcritical fluid extraction can be expensive, which may limit accessibility for small-scale producers or individual users.
- Complexity: The process requires precise control of temperature and pressure, adding a layer of technical complexity compared to simpler extraction methods.
- Limited Solvent Choices: While carbon dioxide is commonly used, it may not be effective for all types of aromatic compounds, particularly polar constituents.

In conclusion, subcritical fluid extraction offers an effective and environmentally friendly option for obtaining aromatic extracts, though its high costs and technical requirements may pose challenges for some users.

CHAPTER THIRTEEN: Supercritical Fluid Extraction

Examples of aromatic extracts obtained via this method: ginger, frankincense, neroli, black cumin, turmeric, vanilla, German chamomile.

Supercritical fluid extraction (SFE) is a modern technique that utilizes fluids at temperatures and pressures above their critical point, where they exhibit properties of both liquids and gases. The most commonly used supercritical fluid is carbon dioxide (CO_2), due to its relatively low critical point, non-toxicity, and non-flammability. By adjusting pressure and temperature, the solvent properties of CO_2 can be tailored to selectively extract aromatic compounds from plant materials. This method produces both select and total extracts. SFE total extracts contain the full spectrum of lipophilic (nonvolatile) and volatile compounds from the plant, while select extracts isolate specific target compounds by adjusting pressure, temperature, and solvent polarity, and more closely resemble distilled essential oils.

Advantages

- High Efficiency: SFE allows for rapid and efficient extraction of aromatic compounds with minimal thermal degradation.
- Tunability: The solvent power of CO_2 can be adjusted by changing pressure and temperature, enabling the selective extraction of target compounds.
- Environmentally Friendly: Using CO_2 as a solvent reduces the environmental impact compared to traditional organic solvents.

- Desirable Aroma: SFE aromatic extracts more closely match the natural aroma of the plant because they preserve delicate volatile compounds that are often degraded by heat in steam distillation or altered by solvent residues in traditional extraction methods.
- Low Residuals: The extraction process leaves minimal residuals, as CO_2 is easily removed from the extract by simply lowering the pressure.

Disadvantages

- High Pressure Equipment: The requirement for high-pressure equipment increases the initial investment and operational costs.
- Limited Solubility: Certain aromatic compounds, particularly polar molecules, may have limited solubility in CO_2, requiring modifications such as adding co-solvents. To extract these, co-solvents (e.g., ethanol, methanol, or water) are often added to modify CO_2 polarity, enhancing the extraction of semi-polar compounds.
- Complexity: The technique demands precise control over extraction conditions, necessitating specialized knowledge and equipment.

Overall, supercritical fluid extraction provides a clean, efficient, and selective method for obtaining aromatic extracts, ideal for high-value applications where the quality and purity of the extract are paramount.

CHAPTER FOURTEEN: CO2 Supercritical Extraction Assisted by Ultrasound

Examples of aromatic extracts obtained via this method: lavender, cinnamon, rosemary, clove, orange.

CO2 supercritical extraction assisted by ultrasound (SC-CO2-UAE) is an advanced technique that combines supercritical fluid extraction (SFE) using carbon dioxide with ultrasonic waves to enhance the extraction of aromatic compounds from plant materials. The ultrasonic energy generates cavitation bubbles in the solvent, leading to improved mass transfer and increased penetration of the solvent into the plant matrix. This synergistic approach allows for more efficient extraction compared to traditional methods.

Advantages

- Enhanced Extraction Efficiency: The ultrasonic waves facilitate the breakdown of plant cell walls, increasing the availability of aromatic compounds and leading to higher yields.
- Selective Extraction: CO2 can be adjusted to selectively target specific compounds, enabling tailored extraction of desired aromatic profiles.
- Lower Temperature: The process minimizes exposure to heat, preserving volatile aromatic compounds and reducing thermal degradation.
- Reduced Solvent Consumption: The combined techniques can lower the required amount of solvent, making the process more sustainable and cost-effective.

Disadvantages

- Equipment Costs: Implementing SC-CO2-UAE requires specialized and potentially high-cost equipment, including pressurized vessels and ultrasonic devices.
- Process Complexity: The integration of ultrasound with supercritical extraction adds complexity to the process, requiring careful control of multiple parameters.
- Limited Availability of Equipment: Not all extraction facilities are equipped with the appropriate technology, limiting widespread adoption.

Generally, CO2 supercritical extraction assisted by ultrasound offers a powerful alternative for extracting aromatic compounds, significantly improving efficiency while maintaining extract quality. However, challenges related to cost and complexity may be barriers to its broader implementation.

CHAPTER FIFTEEN: Cold Extraction with Pressurized Solvents

Examples of aromatic extracts obtained via this method: rose, jasmine, lavender, furanocoumarin-reduced citruses, frankincense, myrrh, thyme, oregano.

Cold extraction with pressurized solvents is an innovative extraction technique that combines lower temperature conditions with the use of pressurized solvents to selectively extract aromatic compounds from plant materials. This method typically employs solvents like ethanol or water, which are maintained at elevated pressures to enhance their extraction efficiency without the detrimental effects of heat.

Advantages

- Preservation of Quality: By operating at lower temperatures, cold extraction helps preserve volatile and thermally sensitive aromatic compounds, ensuring that the quality and organoleptic properties of the extracts are maintained.

- Enhanced Solvent Efficiency: The use of pressure increases the solubility of desirable compounds in the solvent, improving the extraction yield and speed.

- Reduced Energy Consumption: Lower temperatures can lead to energy savings compared to traditional extraction methods that often require heating, making it a cost-effective option.

- Eco-friendly: Using natural solvents reduces harmful environmental impacts associated with petrochemical solvents, making it a safer choice for both producers and consumers.

Disadvantages

- Equipment Costs: The necessity for pressurized systems can involve significant initial investment and maintenance costs.
- Limited Range of Solvents: Not all solvents can be efficiently pressurized, which may restrict the options available for extraction and limit the kinds of aromatic compounds that can be effectively extracted.
- Complexity of Process Control: Managing the pressurization and temperature accurately requires skilled operators and specialized knowledge.

In summary, cold extraction with pressurized solvents presents a viable method for obtaining high-quality aromatic extracts, although its implementation can be hindered by cost and complexity.

CHAPTER SIXTEEN: Pulsed Electric Field (PEF) Extraction

Examples of aromatic extracts obtained via this method: basil, lavender, peppermint.

Pulsed Electric Field (PEF) extraction is an innovative technique that uses short bursts of high-voltage electric pulses to permeabilize cell membranes in plant materials, thereby enhancing the extraction of aromatic compounds. When an electric field is applied, it induces electroporation (a process where tiny holes are made in the cell membrane using an electric current), creating transient pores in the cell membranes. This allows solvents to penetrate more efficiently, facilitating the release of target aromatic compounds into the solvent phase.

Advantages

- Increased Extraction Efficiency: PEF significantly improves mass transfer, leading to higher yields of extracted aromatic compounds in a shorter amount of time compared to traditional methods.

- Low Temperature Processing: The technique minimizes thermal degradation of heat-sensitive compounds by allowing extraction at lower temperatures, preserving the volatile aroma profiles.

- Eco-Friendly: PEF extraction often requires less solvent and energy compared to conventional methods, making it a more sustainable option.

- Rapid Processing: The pulsed nature of the electric field allows for quick extraction cycles, enhancing throughput in industrial applications.

Disadvantages

- Initial Equipment Costs: The technology requires specialized equipment, which can entail significant capital investment and maintenance costs.
- Process Optimization: The effectiveness of PEF extraction depends heavily on parameters such as electric field strength and pulse duration, necessitating careful optimization for specific materials.
- Safety Considerations: High-voltage equipment can pose safety risks if not handled properly, requiring trained personnel to operate.

To conclude, PEF extraction presents a promising method for obtaining high-quality aromatic extracts efficiently while preserving their desirable characteristics, although challenges related to cost and process optimization may limit its widespread adoption.

CHAPTER SEVENTEEN: Enzyme-Assisted Extraction

Examples of aromatic extracts obtained via this method: orange, lemongrass, fennel, vanilla.

Enzyme-assisted extraction (EAE) is a method that utilizes specific enzymes to facilitate the extraction of aromatic compounds from plant materials. By breaking down complex polysaccharides and cell wall structures, enzymes enhance the release of target compounds into the solvent, thereby improving extraction efficiency. This technique can be applied alongside various solvents, including water, ethanol, or organic solvents, and is particularly effective for extracting aromatic oils and flavors.

Advantages

- Higher Yield and Quality: EAE typically results in greater yields of aromatic compounds at lower temperatures, helping to preserve volatile flavors and aromas that might be lost in conventional extraction methods.

- Selective Extraction: The specificity of enzymes allows for targeted extraction of particular compounds, enabling the production of extracts with desired aromatic profiles.

- Environmental Benefits: EAE can reduce the need for harsh solvents and high-energy processes, making it a more sustainable and eco-friendly extraction method.

- Shorter Extraction Time: Enzymatic reactions can significantly shorten extraction times, improving overall process efficiency.

Disadvantages

- Cost of Enzymes: Commercial enzymes can be expensive, which may increase the overall cost of the extraction process.
- Specificity and Stability: Enzyme effectiveness can depend on various factors, including pH and temperature, necessitating careful optimization and control of extraction conditions.
- Potential for Degradation: Improper handling or prolonged exposure can lead to enzyme degradation, which can diminish the effectiveness of the extraction process.
- In conclusion, enzyme-assisted extraction offers a promising avenue for obtaining high-quality aromatic extracts efficiently and sustainably, but challenges related to cost and enzyme stability may need to be addressed for broader application.

CHAPTER EIGHTEEN: Liquid-Liquid Extraction

Examples of aromatic extracts obtained via this method: jasmine, rose, patchouli.

Liquid-liquid extraction (LLE) is a widely used separation technique that separates compounds based on their solubility differences in two immiscible liquids, usually an organic solvent and water. This method is commonly employed for extracting aromatic compounds from plant materials. During the process, the plant material is typically mixed with a solvent, allowing the soluble aromatic compounds to partition into the organic phase, effectively concentrating the desired extracts.

Advantages

- Versatility: LLE can be applied to a wide range of aromatic compounds and is compatible with various solvents, making it adaptable to different extraction scenarios.
- Scalability: The technique can be easily scaled up for industrial applications, accommodating large volumes of material, which makes it suitable for commercial extraction of essential oils and aromas.
- High Purity: LLE can yield high-purity extracts, as the selective nature of the solvent allows for the separation of desired compounds from impurities and unwanted components.

Disadvantages

- Solvent Use: The reliance on organic solvents can pose environmental and safety concerns, including toxicity and potential waste disposal issues.
- Emulsification: LLE can lead to the formation of emulsions—a mixture of two liquids that usually don't mix well, like oil and water, where tiny drops of one liquid are spread out in the other, complicating the separation process and potentially reducing extraction efficiency.
- Time-Consuming: Depending on the solvent and conditions used, LLE may require longer extraction times compared to newer techniques like PEF or enzyme-assisted methods.

In summary, liquid-liquid extraction remains a valuable method for obtaining aromatic extracts, praised for its versatility and scalability, while challenges related to solvent use and efficiency persist.

CHAPTER NINETEEN: Cryogenic Extraction

Examples of aromatic extracts obtained via this method: frankincense, myrrh, sandalwood.

Cryogenic extraction is a highly efficient method that utilizes extremely low temperatures to separate and extract aromatic compounds from plant materials. The process typically involves freezing the biomass using liquid nitrogen or another cryogenic agent, which leads to the rupture of plant cell walls. Once frozen, the material is subjected to a solvent extraction process, allowing aromatic compounds to dissolve into the solvent more effectively than at higher temperatures.

Advantages

- Preservation of Volatile Compounds: Cryogenic extraction helps maintain the integrity of sensitive aromatic constituents, which is crucial for retaining flavor and aroma profiles that might degrade at elevated temperatures.
- High Extraction Efficiency: The method yields high amounts of target compounds due to the enhanced solubility and access provided by the frozen state of the plant material.
- Reduced Solvent Degradation: The low temperatures minimize the risk of solvent degradation, leading to purer extracts with fewer contaminants.

Disadvantages

- High Costs: The need for cryogenic equipment and the use of liquid nitrogen can result in higher operational costs, limiting its feasibility for smaller operations or in regions where cryogenic systems are not accessible.
- Complexity: The process requires specialized knowledge, equipment, and safety measures to handle cryogenic substances safely and effectively.
- Time-Consuming: The processes of freezing and thawing can extend the overall extraction time compared to other methods.

In conclusion, while cryogenic extraction offers significant advantages in terms of quality and efficiency for aromatic extracts, the high costs and complexity associated with the technique may present challenges for widespread use.

CHAPTER TWENTY: Neoteric Solvent Extraction

Examples of aromatic extracts obtained via this method: ylang ylang, sandalwood, rose, eucalyptus, jasmine, rose.

Neoteric solvent extraction refers to the use of novel, non-traditional solvents or solvent systems in the extraction of aromatic compounds from plant materials. These solvents are often more environmentally friendly, have reduced toxicity, and offer superior extraction efficiency compared to conventional organic solvents.

Advantages

- Sustainable and Eco-Friendly: Neoteric solvents are typically derived from renewable sources, biodegradable, or have low environmental impact, making them an attractive alternative to traditional solvents.
- Improved Extraction Efficiency: These solvents can selectively target specific compounds, increasing the yield and purity of extracted aromatics.
- Reduced Energy Consumption: Many neoteric solvents operate at lower temperatures or pressures, which can lead to reduced energy consumption during the extraction process.
- Food Safety and Quality: By minimizing the use of toxic solvents, neoteric extraction can result in higher-quality extracts that meet food safety standards.

Disadvantages

- Higher Costs: The development, production, and use of neoteric solvents can be more expensive than traditional methods due to the need for research and development investments.
- Limited Availability: These solvents may not be widely available or accessible, particularly in regions with limited resources or infrastructure.
- Interoperability and Scalability: Ensuring the compatibility of neoteric solvents with existing equipment and scaling up the process to industrial levels can be challenging.

Overall, neoteric solvent extraction presents a promising approach for the sustainable production of high-quality aromatic extracts, but its adoption is contingent upon addressing the challenges related to cost, availability, and scalability.

CHAPTER TWENTY-ONE: The Future of Aromatic Extracts—Embracing Technology and Sustainability

Technology is advancing, providing additional tools to produce aromatic extracts valuable for health and healing. As our understanding of the complexities associated with essential oil extraction deepens, a diverse range of methods emerges, reflecting the adaptability and ingenuity of modern science. Each technique, whether traditional or cutting-edge, carries its distinct set of advantages and drawbacks, reinforcing the notion that there is no one-size-fits-all approach to extraction. Additionally, the preferred extraction method will be determined by the desired outcome. For instance, if you want a clary sage extract rich in sclareol for therapeutic purposes, you will likely want to use SFE. On the other hand, if your goal is to preserve the sweet, floral notes of ylang ylang for a rich natural perfume, you may want to use neoteric solvent extraction instead. Exploring this landscape of extraction methods reveals not only practical applications but also significant implications for sustainability, efficiency, and quality.

The methods discussed throughout this guide—ranging from steam distillation and cold pressing to advanced techniques like supercritical fluid extraction and enzyme-assisted extraction—demonstrate the breadth of innovation available today. Traditional methods, such as steam distillation, remain popular due to their simplicity and effectiveness, yielding essential oils like lavender, peppermint, and eucalyptus with consistent quality and reliability. The process allows for the preservation of most of the volatile compounds, which is particularly beneficial when working with heat-sensitive botanicals. However, it does transform and even produce some constituents as well. For example, methyl salicylate—found in wintergreen and birch essential oils—is not present in the plant, but produced during distillation, which converts a compound called gaultherin to methyl salicylate. Similarly, chamazulene is produced by conversion of matricine in the plant material to chamazulene. This is why more advanced techniques such as SFE may be used to preserve the gaultherin and matricine.

As demand for diverse and more complex aromatic profiles increases, traditional methods often reveal limitations. For example, hydrodistillation and maceration, while gentle and capable of producing aromatic extracts rich in flavors and aromas, can be time-consuming and less efficient in terms of yield. Alternative methods like molecular distillation and fractional distillation offer higher purity and the potential for custom formulations, yet they require substantial investment and technical expertise.

Emerging extraction methods, such as microwave-assisted and ultrasound-assisted extraction, have proven effective in reducing extraction time and solvent consumption while improving overall yield. These innovations align with increasing consumer awareness around sustainability, as many seek products obtained through eco-friendly practices.

In recognizing the historical relevance of traditional methods, a forward-looking perspective reveals that we can harness nature's potential more effectively with today's advancements. The ability to extract delicate aromas while retaining therapeutic properties is critical in applications across various domains, including food, cosmetics, aromatherapy, and natural therapeutics.

Supercritical Fluid Extraction (SFE), for instance, utilizes carbon dioxide in its supercritical state to produce high-quality extracts with minimal thermal degradation. The versatility of SFE allows for fine-tuning to achieve both select and total extracts, catering to the needs of different industries. This method's ability to achieve high efficiency and preserve aroma integrity has made it particularly desirable for applications where purity matters—such as natural fragrances, dietary supplements, and therapeutic oils.

Cold extraction with pressurized solvents presents another avenue to maintain quality while enhancing efficiency. This method, achieved under pressure, allows for greater solubility of desirable compounds while preventing thermal degradation. Consequently, it has become an eco-conscious choice for extracting aromatic compounds from delicate flowers like rose and jasmine.

The urgency of environmental considerations has heightened awareness around sustainability in extraction methods. Traditional solvents can pose health risks and environmental challenges; thus, the emergence of neoteric solvent extraction using biodegradable and less toxic options is promising. These innovative alternatives align with consumer preferences for safe, sustainable products.

Among these methods is solventless extraction, where processes like hydrodistillation and cold pressing, and inert solvent extraction like supercritical extraction reduce the reliance on potentially harmful solvents entirely. Companies are increasingly exploring water-based and natural solutions, finding ways to minimize waste and energy consumption without compromising quality.

Even more exciting is the integration of technology with nature, such as enzyme-assisted extraction and pulsed electric field extraction. By utilizing enzymes, we can enhance the extraction of specific compounds while reducing the need for harmful solvents. This marriage of biological understanding and technological advancements signals a new era for the botanical extraction industry.

In tandem with sustainability, the increasing demand for transparency in product sourcing has driven the market toward stringent quality control measures. Consumers want to know the origins of their products and the methods by which they were extracted. This shift has prompted greater regulatory scrutiny and standardization across the industry, resulting in improved traceability and adherence to safety standards.

Because extraction methods directly influence the aromatic profiles of essential oils and extracts, producers are encouraged to adopt practices that not only yield better results but also prioritize the overall safety and efficacy of their products. Enhanced analytical techniques, such as gas chromatography-mass spectrometry (GC-MS), offer insights into the chemical composition of essential oils, ensuring quality and authenticity. Comparison of this data across extraction methods helps guide producers to the most preferred method for each botanical extracted from. Doing so ensures that the right constituents in the right ratios are present, and the less desirable constituents are minimized.

Another major reason we need detailed analytical methods beyond GC-MS is the adulteration present in the industry. The essential oil industry is plagued by a pervasive problem of adulteration, where unscrupulous manufacturers knowingly alter essential oils to increase profits, compromising their quality, safety, and efficacy. This rampant adulteration poses significant risks to consumers, as tampered oils can cause adverse health reactions, trigger allergies, or even lead to toxicity. At the very least, an adulterated oil cannot be relied upon to produce the same benefits as the body will respond differently to a synthetic chemical than a natural chemical.

The lack of transparency and regulation in the industry exacerbates the issue, making it crucial to have qualified chemists who can detect and identify these adulterated oils. Through advanced analytical techniques such as GC-MS, chiral GC-MS, and nuclear magnetic resonance (NMR) spectroscopy, skilled chemists can uncover the presence of synthetic additives, solvents, or other foreign substances that may have been added to the oils. Without proper testing and authentication, consumers may unknowingly expose themselves to harmful substances, highlighting the urgent need for stringent quality control measures and expert chemical analysis to ensure the authenticity and safety of essential oils. The consequences of ignoring this issue can be severe, ranging from skin irritation and respiratory problems to more serious health conditions, underscoring the importance of vigilance and expertise in protecting the integrity of the essential oil industry.

The importance of quality control cannot be overstated, as inconsistencies in extraction methods can lead to variations that may affect consumer trust and brand reputation. This focus on quality helps producers align their operations with contemporary consumer values, fostering brand loyalty rather than just a transaction-based relationship.

Looking ahead, the landscape of aromatic extraction is poised for remarkable transformation. As we continue to enrich our understanding of plant phytochemistry, we can expect to witness the emergence of hybrid techniques that combine multiple extraction methods for enhanced results. This synergy between traditional wisdom and modern technology empowers producers to craft products with greater complexity and specificity than ever before.

Furthermore, collaborative efforts within the industry can lead to the development of best practices and shared knowledge, ultimately benefiting smaller producers and large corporations alike. Establishing industry standards for sustainable practices and responsible sourcing of raw materials will stimulate healthy competition and drive innovation without compromising ethical considerations.

The rise of consumer education and advocacy also plays a significant role in shaping the future of aromatic extraction. As individuals increasingly prioritize natural products, it falls upon producers to respond not only by refining their extraction methods but also by addressing the broader ecological implications of their practices. This shift presents an opportunity to explore niche markets and respond to evolving consumer demands while contributing to environmental preservation.

The quest for higher-quality aromatic extracts cannot simply focus on yield; it must encompass a holistic perspective, recognizing the interconnectedness of human health, environmental sustainability, and economic viability. Embracing sustainable practices, ethical sourcing, and innovative extraction methods enhances product quality and fosters a greater sense of corporate responsibility in the industry.

As the chapter on aromatic extraction techniques closes, it is essential to open a new one focused on continuous improvement, learning, and adapting to changing realities. Both producers and consumers share a collective responsibility to harness the gifts of nature while treating them with respect and care, ensuring the excavation of aromatic treasures does not come at excess cost to the planet.

In conclusion, the future of aromatic extract production is bright and promising. With ongoing advancements in technology and an unyielding commitment to sustainability, we stand at the horizon of an exciting new chapter in the world of essential oils and aromatic extracts. By prioritizing innovation, quality, and environmental stewardship, we can ensure that these aromatic treasures continue to enrich our lives for generations to come. The journey of extraction is far from over; it is an ever-evolving narrative that reflects our growing understanding of nature and our role in preserving its aromatic gifts.

Discover More by Dr. Scott A. Johnson

The Most Trusted and Respected Natural Health Books!

- *Medicinal Essential Oils (Second Edition): The Science and Practice of Evidence-Based Essential Oil Therapy*
- *Co2 Aromatics, Medicinal Herbs, and Targeted Nutraceuticals for Healing and Greater Wellness*
- *Evidence-Based Essential Oil Therapy: The Ultimate Guide to the Therapeutic and Clinical Application of Essential Oils*
- *Surviving When Modern Medicine Fails: A definitive Guide to Essential Oils That Could Save Your Life During a Crisis*
- *Como sobrevivir cuando la medicina moderna falla – tercera edicion: La guia absoluta sobre los aceites esenciales que podrian salvarle la vida durante un suceso critico* (Spanish Edition)
- Livro Aromaterapia e Momentos de Crise (Portuguese Edition)
- *Heal the Gut, Heal the Immune System*
- *What Big Pharma Doesn't Want You to Know About Essential Oils*
- *Synergy, It's an Essential Oil Thing: Revealing the Science of Essential Oil Synergy with Cells, Genes, and Human Health*
- *The Waterfall Technique: A Whole-Body Essential Oil and Restorative Touch Experience to Rejuvenate, Restore, and Optimize*

- *Monographs of Rare and Exotic Essential Oils and Absolutes: Exploring the Past to Discover the Future of Medicine*
- *End-of-Life Care with Essential Oils: Your Guide to Compassionate Care for Loved Ones and Their Caregivers*
- *Beating Histamine Intolerance Naturally*
- *The Endocannabinoid System and Cannabis: The Perfect Partnership for Self-Regulation and Healing*
- *Cannabis E O Sistema Endocanabinóide*
- *Improving Human Health through the Healing Power of Medicinal Mushrooms*
- *Beating Ankylosing Spondylitis Naturally*
- *Beating Autoimmune Thyroid Conditions Naturally*
- *Beating ADHD Naturally*
- *The Vaccine Book Parents Deserve: Empowering and Reliable Evidence to Make an Informed Choice for Your Children*
- *Disruptive Mood Dysregulation Disorder: An Empowering Integrative Guide for Parents*
- *The Doctor's Guide to Surviving When Modern Medicine Fails: The Ultimate Natural Medicine Guide to Preventing Disease and Living Longer*
- *TransformWise: Your Complete Guide to a Wise Body Transformation Kindle Edition*
- *Nutrition and Integrative Medicine: A Primer for Clinicians* – 1st Edition (Dr. Johnson wrote the chapter on essential oils in this book)
- *The Word of Wisdom: Discovering the LDS Code of Health*

- *Regaining Humanity: 15 Essential Character Traits to Unplug from Rule by the Elites*
- *Jeremy's Christmas Journey* (Book & CD)

www.ingramcontent.com/pod-product-compliance
Lightning Source LLC
Chambersburg PA
CBHW060521280326
41933CB00014B/3061